Rising Parent Media, LLC
 Published 2018
Printed in the United States of America

22 21 20 19 18 1 2 3 4

ISBN: 978-0-9987312-4-7

The paper used in this publication meets the minimum requirements of the American National Standard for Information Sciences—Permanence of Paper for Printed Library Materials, ANSI Z39.48-1992.

FOR GREAT RESOURCES AND INFORMATION, FOLLOW US ON OUR SOCIAL MEDIA OUTLETS:

Facebook: www.facebook.com/educateempowerkids/
Twitter: @EduEmpowerKids
Pinterest: pinterest.com/educateempower/
Instagram: Eduempowerkids

www.educateempowerkids.org

Special thanks to the Center for Media Literacy, Common Sense Media, Media Literacy Now, Media Smarts, and the National Association for Media Literacy Education for their wealth of knowledge and experience and their decades of education and advocacy. Their resources inspired us to create this book.

PETRA'S POWER TO SEE
A MEDIA LITERACY ADVENTURE

Educate and Empower Kids would like to acknowledge the following people who contributed time, talents, and energy to this publication.

STORY BY:
Dina Alexander, MS

EDITED BY:
Melody Bergman
Tina Mattsson
Maren Warner
Jenny Webb, MA

DESIGN AND ILLUSTRATION BY:
Jera Mehrdad
Jenny Webb, MA

SPECIAL THANKS TO:
Kyle Roberts, MA
Cliff Park, MBA
Byran Korth, PhD

FAUX FUR
KEEP YOURSELF WARM AND FABULOUS
FIND YOUR NEW HIGH!
GREAT WHEEL BIKES

INTRODUCTION

We are surrounded by messages; many of them are beautiful! Some inspire us to learn and grow. But some messages are useless or even harmful.

Understanding messages takes practice. We can learn to deconstruct media messages so we can SEE all their parts. As you read this book, you will see that any media can be deconstructed with simple questions.

This book has lots of information. Feel free to read and talk about just a few pages at a time.

Use your ***POWER TO SEE*** in the practice sections following each chapter. Remember, the questions found there do not necessarily have right or wrong answers. They are meant to encourage conversation and understanding.

New Words: We will be introducing some new words throughout the book. These words will be in underlined and found at the end of the book, in the Glossary. and found at the end of the book, in the Glossary.

STOP
SAVE
LIVES

Made in the USA
Middletown, DE
28 November 2019

79597459R00040

CHAPTER 1: WHAT IS MEDIA?

Petra and Dad boarded a busy train headed to the city for an adventure.

Inside the train, Petra could hear the pings of another child's video game, feel the thump of music leaking from someone's headphones, and see a YouTube video playing on the man's tablet across from her.

Next to her, Dad was looking through his Facebook feed.

Outside the train, there were messages everywhere they looked.

There were billboards, movie posters, digital signs, window displays, and more.

"What do you think of all these messages around us, Petra?" Dad asked.

"What messages?" Petra replied, feeling confused.

FAUX FUR
KEEP YOURSELF
WARM AND
FABULOUS
CREATURE COMFORTS
Velvet
perfection all day
New
Ultra
Rich Color
?!?!?

"Look around," said Dad. "See the billboards? See that lady's magazine and that man's video?"

"Yes, I see them" Petra said, still feeling puzzled.

"Those are all sending messages, aren't they?" Dad asked.

Petra thought for a moment, her eyes widening. "Wow, I hadn't thought about it like that before. They are all sending messages!"

"Did you know all these messages have a name?" asked Dad. "They are all types of media."

"What's media?" Petra asked.

"MEDIA IS A WAY TO SEND MESSAGES"

replied Dad.

"It's also a way to talk to a lot of people at once."

"When I was little, media was TV, radio, books, and newspapers. Now it's so much more than that."

"Like what?" Petra asked.

"Like the music playing from that kid's headphones, or the game on that phone. Everything you see on the internet is media too: every ad, every website, every app."

"Wow," Petra was amazed.
"Media is everywhere."

LET'S PRACTICE WITH THE POWER TO SEE

CHAPTER 1: WHAT IS MEDIA?

There are messages everywhere we look. At home, at school, in our community.

They are all around us!

QUESTIONS TO THINK ABOUT:

If "media is everywhere," how does it affect our choices?

Does it shape our thoughts and beliefs?

We often need media in our schoolwork and homework. Is there an ideal amount of time we should spend with media?

How much is too much media in a day? A week?

ACTIVITY:

COUNTING THE MESSAGES IN OUR DAILY LIVES

How often do you see ads? Try counting all the ads you see in one day.

What kinds of ads stand out to you, such as ads for toys, travel, fast food, or clothes?

How many other messages will you see in one day? Some places you might see media are online, TV, movies, books, games, social media, magazines, and store fronts.

What other messages do you see in your community such as from churches, community centers, libraries, government buildings, etc.?

Where do you see the most messages in a day? On TV? On a phone? Computer?

CHAPTER 2:

DECONSTRUCTING MEDIA

Looking around the train, Petra examined a picture on the wall above the seats.

"Hey Dad, look," Petra said, tugging at his arm. "There's some media right over there."

"Good eye, Petra." Dad said, following Petra's stare. "It's an advertisement. What's going on in that ad?"

"Hmm. Well, a boy is jumping off a ramp on a really cool bike," replied Petra.

"You're right!" Dad exclaimed. "What's the ad trying to tell you?"

"It's saying this is a really fun bike kids can do jumps with," said Petra.

ADVERTISEMENT:

A picture, video, or words that try to persuade people to buy a product or service. The words "ad" or "ads" are short for advertisement.

FIND YOUR
NEW HIGH!

"Let's look a little deeper at it. Is there a hidden message?" Dad asked.

"What do you mean?" asked Petra.

"Well, most advertisements have <u>underlying</u> or hidden messages that you don't see at first," explained Dad. "You usually have to stop and look a little harder."

"Oh, I get it. Like this ad is also saying if you get this bike, then you will have fun and be cool like this kid," Petra said.

"Exactly," said Dad. "So, can you guess why this ad was created?"

"Sure," said Petra. "It was made for kids, so we will want to buy this bike."

Dad smiled. "Does it make you want a new bike?"

"Maybe," Petra replied. "But 'yes' if it comes in orange, and my friends have one too."

"We just deconstructed that ad," said Dad.

"What does '<u>deconstructed</u>' mean?" Petra asked, confused.

"We broke it into pieces," Dad explained. "We talked about its overall and hidden messages. Then we figured out why it was made and who it was made for."

Petra looked closer at the advertisement.

"You see it differently now, don't you?" Dad asked.

"Definitely!" exclaimed Petra. "When we deconstruct ads," Dad said, "we give ourselves power. We have the *POWER TO SEE*. With the *POWER TO SEE*, we can see media for what it really is."

DECONSTRUCT MEDIA:

To break down an ad, movie, commercial, song, or other media into pieces to see what the real message is.

START
START
STRON
STRONGE
XCUSES.
MONTHLY RATES.

Getting off the train, Petra saw another ad that got her attention.

"I've seen ads from this gym all over the place! Let's deconstruct this one too, Dad," said Petra.

"Okay," Dad said. "First, what got your attention?"

"His huge muscles and his face. He looks crazy excited about his workout," Petra replied.

"So what's the message in this ad?" asked Dad.

"Well, he looks pretty tough," Petra replied. "I think they are trying to say if you work out at their gym, you'll be super strong like this guy."

CHAPTER 2: DECONSTRUCTING MEDIA

An advertisement can be a very powerful, convincing form of media. Each person sees and hears ads and media in different ways.

To give yourself the ***POWER TO SEE*** different media, you need to DECONSTRUCT the messages.

WHAT DO YOU SEE?

What methods were used in this ad to get our attention?

What is the underlying or hidden message?

What do you like about it?

What do you dislike?

What do the ad designers want us to do?

How might your friends feel about this ad?

What would your parents think about it?

How about your grandparents?

WHAT DO YOU SEE?

What is the overall message?

Why was this ad made?

What did they use to sell this product? Some tools advertisers use to make their product more appealing can include the objects in the image, words, and font styles.

How does this advertisement make you feel? Does it make you want to eat a slice of pizza?

CHAPTER 3: TYPES OF MEDIA

After stopping for lunch, Petra and Dad finally arrived at their destination, a wonderful old theater, to catch a movie.

Afterward, they discussed the show.

"I loved it!" Petra announced. "It was funny, the music was cool, and that was the best ending ever. Wait a second. Are movies 'media' too?"

"Definitely," said Dad. "So are comic books, cartoons, plays, short stories, and YouTube videos."

"Woah! I'm beginning to SEE media all around us," Petra said.

$1 IS ALL IT TAKES TO VACCINATE THIS CHILD
DONATE TODAY
SAVE OUR CHILDREN FOUNDATION
SOCF.ORG
PERFECT HAIR
OOL VIBES
TREAT YOURSELF!
Sweet Tweets Bakery
HEROMAN
THE MOVIE

CHAPTER 3: TYPES OF MEDIA

Having the *POWER TO SEE* takes practice! Another name for this power is media literacy.

Some media messages are information you need to know like history, science, or how to get along with other people.

MEDIA LITERACY:

A way to study and understand MESSAGES in different kinds of media such as books, movies, music, news stories, online ads, or school essays.

Some media messages tell you what an advertiser or company thinks you should eat, how you should look, what you should wear, what you should think, and how you should act.

You can stop and deconstruct ANY media in your life to find the real messages you get from media.

YOU CAN ASK YOURSELF:

Is this commercial truthful?

What did this book tell me?

What did this TV show not tell me?

How does this video make me feel?

Why was this app created?

ZABAR'S
ZABAR'S
CELEBRATE... WITH
ZABAR'S
Caviar
Visit us at Zabars.com
MEZZANINE
ZABAR'S
SENIOR CITIZEN
10% DISCOUNT
EVERYDAY
NOT AN ENTRANCE
NOT AN ENTRANCE
Help Wanted
ZAGAT GUIDES
ZABAR'S
Celebrating our 80th Anniversary
1934–2014
ZABAR'S
Celebrating our 80th Anniversary!
1934–2014

Perfection.
All day.
Just like
you.
velvet lipstick
velvetlipstick.com

CHAPTER 4: MEDIA ILLUSIONS

Turning a corner, Dad pointed out a huge billboard across the street.

"What do you think about this ad, Petra?" Dad asked. "Do you think the creators used a computer program to change this image to make her more 'perfect'?"

"Can they do that?" wondered Petra.

"Of course! Look closely at her. She was already pretty, but the designer removed the natural lines on her face. She doesn't even have any pores on her face like everyone has."

"That's weird. Why did they do that?" asked Petra.

"Most advertisers think if they show perfect-looking people with a perfect-looking product, we will want to buy the product," Dad explained.

"And it's not just lipstick ads. They alter ads for toys, cars, clothes, food—almost everything."

THINK:

The ad says "Perfection. All day." Do you think wearing this lipstick will make you look perfect?

"Is that honest? To change the way people and products look?" Petra demanded.

"Well, it's not an outright lie," explained Dad. "But most people feel it's not 100% honest."

"It just doesn't seem right," Petra said, fired up. "They're trying to trick us into buying their

THINK:

Do you think the people who made this did anything to make her look more perfect?

stuff. They want us to think their products will make us happy or perfect like the people in the ads. But that's just not true!" said Petra.

"Wow Petra, you really are starting to SEE," said Dad.

CHAPTER 4: MEDIA ILLUSIONS

Computer enhancements are very common in media.

DISCUSS THE FOLLOWING QUESTIONS:

Is it a good idea for advertisers to alter photos and videos?

Do people in beauty ads look like people you know?

What feelings do the advertisers want you to feel when they show beauty or fitness products?

How do beauty and fitness ads make you feel?

Is Petra right? Do advertisers want us to think our lives will be better if we buy what they are selling?

On the left side of this photo, we see a model with no makeup and her natural hair.

On the right, we see the same model with makeup, hair done, and Photoshop enhancements to take away any blemishes or "imperfections." This is very common in advertising, videos, and other media.

Is one of these pictures better than the other? Why or why not?

SOCIAL MEDIA:

Websites and apps that are used by people over the age of 13 to share information and develop personal and professional relationships.

CHAPTER 5: SOCIAL MEDIA

Petra and her dad were back on the train heading home.

"Is that media too, Dad?" asked Petra.

"It hadn't occurred to me before, but it is media—social media," said Dad.

Petra looked at Dad's phone where she saw a photo of their neighbor. Thinking hard, Petra asked, "So does she have a message?"

"Definitely," Dad explained. "Whenever we post on Facebook, Instagram, or Twitter, we are sharing a message."

"Cool! So kids can create media too!" Petra realized.

CHAPTER 5: SOCIAL MEDIA

You can use your **POWER TO SEE** to DECONSTRUCT social media too.

WHAT DO YOU SEE IN THIS POST?

What are some of the messages your friends and family share on social media?

Do their posts have underlying or hidden messages?

Is what they post true?

How do you respond to their posts and photos?

Social media is powerful! We can use this power to reach hundreds or thousands of people. We can choose to tear people down or lift them up.

POSITIVE THINGS WE CAN DO ON SOCIAL MEDIA:

SINCERELY COMPLIMENT OTHERS

RESPOND KINDLY WHEN YOU DISAGREE WITH SOMEONE

BE TRUTHFUL

SHARE HELPFUL NEWS AND INFORMATION

SHOW EMPATHY WITH THOSE WHO SHARE DIFFICULTIES IN THEIR LIVES

SAY ONLY THINGS YOU ARE WILLING TO SAY IN PERSON

#THINKBEFOREYOUPOST #USETECH4GOOD

ASK YOURSELF:

When I post on social media, am I being clear in my message? • Do I really understand the issue or concern I am posting about? • Am I being true to myself? Am I being authentic? • Is my message helpful? • Is my message kind? • How will my message be received? • Is this something I want on the internet FOREVER?

CHAPTER 6:

REAL NEWS VS. FAKE NEWS

Dad switched social media feeds and clicked on a news story.

"This is getting out of hand!" Dad exclaimed, rolling his eyes.

"What's getting out of hand? asked Petra

"These fake news stories," said Dad.

Petra looked down and read the headline. "What!? News can be fake?"

"Yes. See this story?" asked Dad. "It looks like any other news article except it's full of half-truths.

"How are we supposed to know what's real and what's fake?" Petra asked.

"We can practice our *POWER TO SEE* and deconstruct news too," replied Dad.

"First, we need to read more than the headline. Then we need to stop and think about what we just read. Sometimes 'news' articles make big promises about getting rich quick, losing weight with very little effort, or other untrue claims. If it sounds too good to be true, it probably is."

"And Petra, if there is anything you don't understand, or if something doesn't feel right, you can always ask me or Mom."

LET'S PRACTICE WITH THE POWER TO SEE

CHAPTER 6: REAL NEWS VS. FAKE NEWS

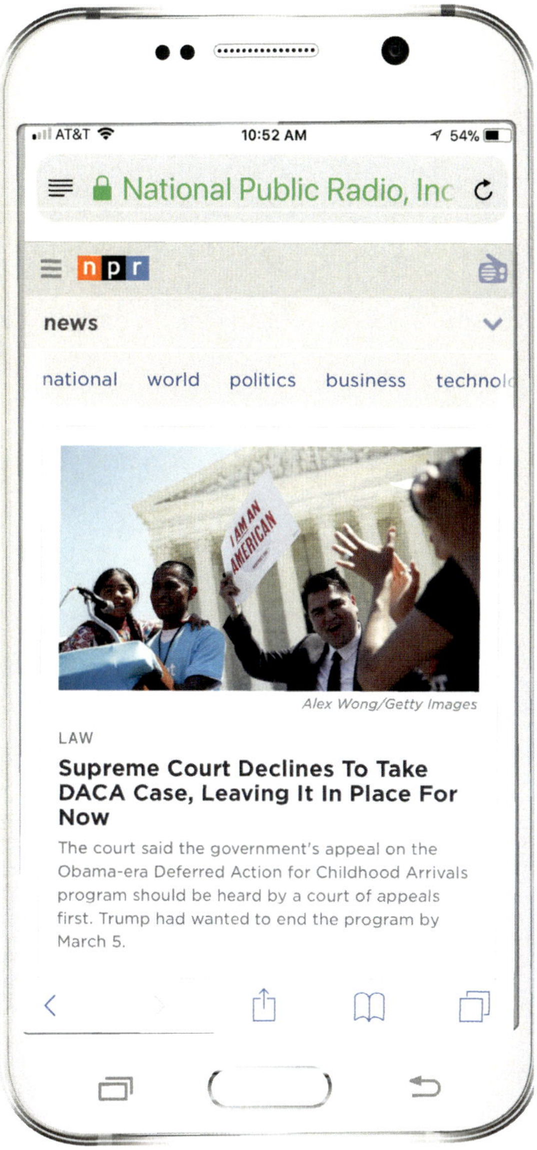

We need to be reading with our eyes wide open!

Pick a current event/ news story to discuss. Read the headline and the article, hopefully with a parent or teacher. The following questions will help you know if the article is real or fake.

DISCUSS THESE QUESTIONS:

Who is the author? Is this a real person?

If an "expert" is quoted in the story, can he or she be found? Are they really a doctor, a political expert, etc.?

What is the source of this story? What website or news agency published the story?

Are the other news stories on this website believable?

QUESTIONS FOR OLDER KIDS:

Think about how you found the news story.

Did it show up in your social media feed? Was the story "promoted" by a website or is it "paid content?"

Who funds or pays for this website/publication?

Is the article a satire or a joke?

Do you have a personal bias that helps you believe or disbelieve the article?

Pay special attention to the last question. Sometimes our prejudices are more powerful than the actual news we consume. If you are still in doubt about the truthfulness of certain news, check various sources for information, including your parents and teachers. These websites are also useful for finding accurate information: factcheck.org and snopes.com.

CHAPTER 7: MEDIA TO AVOID

Nearing home, Dad put away his phone. "Petra," he said seriously, "There is some media we should stay away from."

"Like what?" she asked, surprised.

"Well, some media is violent," Dad said.

"Mom told me if a video game or movie has too much hitting, name-calling, or worse, it's unhealthy to watch," replied Petra.

"She's right. It's very hard to 'un-see' or 'un-hear' TV shows, songs, or other media," explained Dad. "And worse, watching too much violence can also affect the way we think about and the way we SEE real people."

"Is there any other media I should stay away from?" Petra asked.

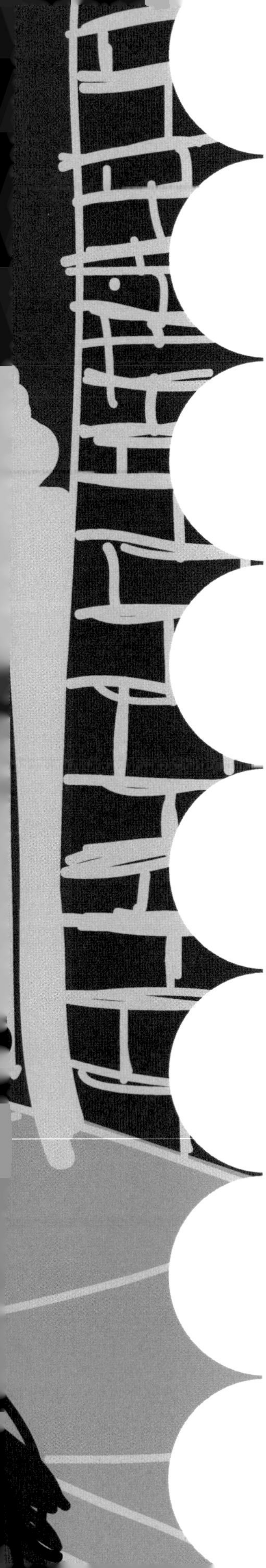

"Yes, there is," said Dad sadly. "Since computers and cellphones are everywhere, you need to be careful about pornography."

"What's pornography?" Petra asked, confused.

"Pornography is pictures or videos of people with little or no clothing on. Usually there is sexual behavior. It shows private actions in order to make money."

Petra was a bit shocked at hearing about this type of media.

"When you do see it, please tell me or Mom what you saw," said Dad. "We can help you make sense of it, and we can help you make sense of your feelings, too."

"I will Dad," Petra said firmly.

"It's natural to be curious about sex," Dad said. "But the messages in pornography are some of the worst and most incorrect sources of information.

You're going to come across violent media and pornography, but you can choose now to never seek it out."

"Pay attention to your feelings when you watch a show or play a video game. Does it give you sad or icky feelings? If it does, get away from it!"

LET'S PRACTICE WITH THE POWER TO SEE

CHAPTER 7: MEDIA TO AVOID

R-rated movies and games rated Mature or Adult should not be used by kids. Heavy exposure to violent media can lead to desensitization. Just because a friend is allowed to play violent games or watch violent movies does NOT mean they are healthy for you.

DESENSITIZATION:

When a person becomes less sensitive or feels no shock or distress at scenes of cruelty, violence, or suffering because of heavy exposure to such images.

DISCUSS THESE QUESTIONS:

What is violence?

How do we know if a TV show, movie, or game is too violent?

What should you do if you are at a friend's house, and he or she wants to play a violent video game?

With your parents, discuss possible places you might be exposed to porn or violent media, such as on a home computer while doing homework, on the school bus, on a friend's phone, or social media, etc. You need a plan for dealing with exposure. We recommend talking about and using the RUN plan. If you are exposed:

R-U-N PLAN FOR KIDS

RECOGNIZE WHAT YOU'VE SEEN AND GET AWAY FROM IT.

UNDERSTAND WHAT YOU'VE SEEN AND TALK ABOUT HOW IT MADE YOU FEEL WITH A TRUSTED ADULT.

NEVER SEEK IT OUT AGAIN.

DISCUSS THESE QUESTIONS:

What is a simple way of telling your friend you don't want to look at something inappropriate on their phone?

How can you tell your parents when you have seen something pornographic or violent?

Are there places, online and offline, you can avoid so you don't see violent or pornographic media?

CHAPTER 8: THE POWER TO SEE

As the train pulled into the station, Dad wanted to share one more thought with Petra.

"We've talked a lot about different media today, but I want you to understand the messages you see everyday and prepare you for other media you'll see in the future," Dad said. "Media isn't bad, though. It can be wonderful!"

"Yeah, like the great movie we saw today," said Petra.

"Or the amazing books we read together and the beautiful music we love," Dad added. "We can learn and enjoy so much from media, especially if we are prepared to SEE the real messages inside it. We can have the ***THE POWER TO SEE***."

Petra smiled, "I'm ready Dad. I'm ready to SEE."

EXTRA PRACTICE WITH

THE POWER TO SEE

Any media can be deconstructed with simple questions. Use your ***POWER TO SEE*** to discuss various media and its effect on our lives. Remember, these questions do not necessarily have right or wrong answers. They are meant to spur conversation and understanding.

WEBSITES

Go to a website with your parent or teacher.

DISCUSS THESE QUESTIONS TOGETHER:

What do you see on the homepage?

What do you like about it? What do you dislike?

What did the web designer do to get our attention?

Who is supposed to use this website (kids, adults, teens)?

What is the point of this website? To sell products? To educate? To entertain?

What does this website tell me about how other people think or live?

What lifestyle and values are represented here?

What lifestyles and values are left out?

MUSIC

Play a favorite song, paying attention to the beat, melody, and rhythm. Look up the lyrics online, and read them.

DISCUSS THESE QUESTIONS:

Who created this message?

What is the overall message?

What creative techniques are used to get my attention?
Some examples may include the rhythm, beat, lyrics, etc.

What lifestyle and point of view are expressed?

What lifestyles and points of view are left out?

How might other people understand this message differently than me?

Why was this music created?

MOVIE, TV, OR SHORT VIDEOS

Watch a movie or television show together, or discuss one you have recently viewed.

What did you think of the hero?

What do you think was the message the creators wanted us to see and hear?

What creative techniques were used to keep our attention? Some examples may include music, actors, dialogue, jokes, special effects, etc.

Is this movie good for kids? Why or why not?

Did you notice any stereotypes?

STEREOTYPE:

A generalized, oversimplified image or idea about a group of people.

Was there anything controversial in this movie?

MEDIA RATINGS

Knowing what different media ratings mean can help you choose what is appropriate for you. The following are TV ratings (on the left) and Game ratings (on the right).

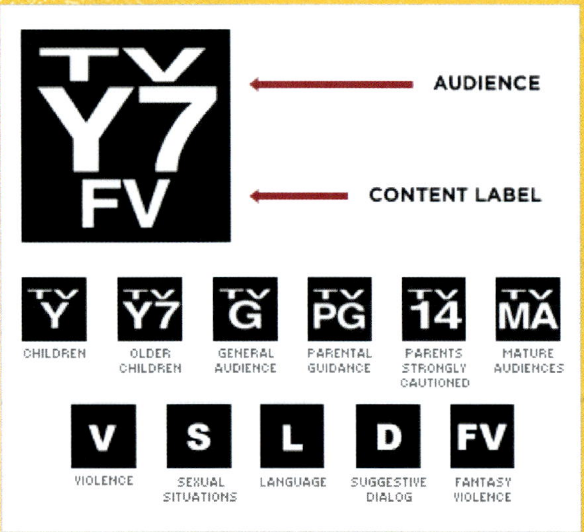
TV Y7 FV
AUDIENCE
CONTENT LABEL
TV Y
CHILDREN
TV Y7
OLDER CHILDREN
TV G
GENERAL AUDIENCE
TV PG
PARENTAL GUIDANCE
TV 14
PARENTS STRONGLY CAUTIONED
TV MA
MATURE AUDIENCES
V
VIOLENCE
S
SEXUAL SITUATIONS
L
LANGUAGE
D
SUGGESTIVE DIALOG
FV
FANTASY VIOLENCE

EARLY CHILDHOOD
eC
CONTENT RATED BY ESRB
EVERYONE
E
CONTENT RATED BY ESRB
EVERYONE 10+
E 10+
CONTENT RATED BY ESRB
TEEN
T
CONTENT RATED BY ESRB
MATURE 17+
M
CONTENT RATED BY ESRB
ADULTS ONLY 18+
AO
CONTENT RATED BY ESRB
RATING PENDING
RP
CONTENT RATED BY ESRB

VIDEO GAMES

Video Games can be really fun to deconstruct!

THINK ABOUT THE GAMES YOU PLAY:

What are some of the overall messages in the games you play?

Do they have any underlying or hidden messages?

How do video games get our attention?

What are your favorite games teaching you?

Why do people create video games?

How do you decide what your favorite games are? Do you like the characters, graphics, challenges, playing with other people, or game goals?

How do games that are rated for "Everyone," "Teen," "Mature," and "Adult" differ?

Since there are no real people in the game, is any amount of violence okay?

If it's NOT okay for kids to play, is it okay for adults to play certain games?

How much time gaming is too much? You decide.

MORE GREAT DISCUSSIONS

What is your favorite type of media?

How does media affect our choices or shape our beliefs, political views, and attitudes?

Which forms of media affect you the most?

Different people can experience the same piece of media differently. Something that you may like or think is funny may not be appealing to someone else. How can you be kind and empathetic to others when you create media?

Is it okay to have a different opinion about music, movies, books, etc. than your parents?

How can we truly understand the media messages we see, hear, and experience around us?

Most media is made for the purpose of making money. Is that wrong?

With media we can change the world for the better or for the worse. What can you do with social media or other media to make the world a better place?

Many people say "sex sells" or "violence sells." What do they mean by this? Can you think of any examples of this?

How can you avoid the worst of media?

How can you seek out the best?

What kinds of messages do you send in the media you create? Are you clear and honest?

HOUSEHOLD MEDIA GUIDELINE

In a family meeting, make a personalized, detailed guide and agreement—a standard—within your family to determine what types of media, devices, timeframes, and protection tools will be used. Have a discussion with your kids and teenagers; talk about the "why's," not just the "no's."

WHAT TO INCLUDE: Each family has to determine what their media standard will contain based on their individual circumstances and needs. However, we suggest including the following:

TAKING INVENTORY

Our family's internet-enabled devices: ______________________

____

____

How many hours of screen time are we each using each day?

(name): __________________ **(hours):** ________

(name): __________________ **(hours):** ________

(name): __________________ **(hours):** ________

(name): __________________ **(hours):** ________

Where do we keep our devices? ______________________

____

What video games/TV shows/social media are we using? ______

____

____

____

SETTING THE STANDARD

What kind of media do we allow? ______________________________
____
____

What are our time limits for screen-time? ____________________
____
____

What are our rules for social media? _________________________
____
____

What filters, parental controls, and/or accountability software will we use? ____
____

Accountability and consequences: Your guideline should address these issues in order to work. Discuss as a family how each of you will be accountable to stick to the guideline and what the consequences will be for breaking the rules. Make accountability part of the plan, but make it as difficult as possible to break the rules by installing filters and restrictions. Talk about expectations for media usage outside your home and what specific websites, social media accounts, games, movies, ratings, etc. are not allowed.

Adapted from Fight the New Drug's Family Media Standard. Family Media Standard. (2015). Retrieved August 25, 2015.

GLOSSARY:

ADVERTISEMENT: A picture, video, or words that try to persuade people to buy a product or service.

BIAS: A personal view, attitude, or tendency.

CONTROVERSIAL: Topics that people discuss and debate because they have strong, differing opinions. Sometimes these debates can be contentious.

DECONSTRUCT MEDIA: To break down an ad, movie, commercial, song, or other media into pieces to see what the real message is.

DESENSITIZATION: When a person becomes less sensitive or feels no shock or distress at scenes of cruelty, violence, or suffering because of heavy exposure to such images.

EMPATHY: The ability to understand and share the feelings of another person.

FAKE NEWS: A story that looks like any other news article, except that it's not true or it's full of half-truths. A lot of people, adults included, can't tell the difference.

INQUIRY: Asking for information, seeking for truth or knowledge. An investigation. This is vital in understanding media.

LIFESTYLE: The way a person, group, or culture lives. In advertising, products are shown in a desirable lifestyle.

MEDIA: A way to send messages or speak to many people. Media can also influence many people. Radio, television, newspapers, and the internet are examples.

MEDIA LITERACY: The POWER TO SEE. The ability to study, understand, and create messages in various media such as books, social media posts and

photos, movies, games, music, news stories, online ads, blog posts, school essays, etc.

PORNOGRAPHY: Pictures or videos of people with little or no clothing. It shows private actions in order to make money. Usually there is sexual behavior in it.

SATIRE: The use of humor, irony, exaggeration, or ridicule to expose and criticize people's stupidity or vices, particularly in the context of contemporary politics and other topical issues.

SOCIAL MEDIA: Websites and apps that people use to share information and develop personal and professional relationships. Facebook, Google+, Instagram, Pinterest, Snapchat, and Twitter are examples of social media.

SOURCE: A book, person, statement, etc. that is supplying information.

STEREOTYPE: A generalized, oversimplified image or idea about a group of people.

TARGET AUDIENCE: The group of people professionals are talking to when they create media. The creators always have a specific audience and a specific purpose in mind.

UNDERLYING MESSAGE: A message that is only revealed after close inspection of an advertisement, song, story, or other media. These messages are not obvious and are often hidden or masked in some way. Usually, discovering them requires deep thought and critical thinking.

VALUES: A person's beliefs, personal rules, and standards.

VIOLENCE: The use of physical force so as to injure, abuse, damage, or destroy.

CITATIONS:

Advertisement Meaning in the Cambridge English Dictionary. (n.d.). Retrieved March 31, 2017, from http://dictionary.cambridge.org/dictionary/english/advertisement

Advertising's toxic effect on eating and body image. (2015, March 18). Retrieved March 26, 2017, from https://www.hsph.harvard.edu/news/features/advertisings-toxic-effect-on-eating-and-body-image/

Alexander, D., M, & Mehrdad, J. (2016). How to talk to your kids about pornography. United states: Rising Parent Media.

Carson, J. (2017, February 08). What is fake news? Its origins and how it grew in 2016. Retrieved March 25, 2017, from http://www.telegraph.co.uk/technology/0/fake-news-origins-grew-2016/

Co-Viewing With Your Kids - Tip Sheet. (n.d.). Retrieved March 27, 2017, from http://mediasmarts.ca/teacher-resources/co-viewing-your-kids-tip-sheet

Dachis, A. (2011, July 25). How Advertising Manipulates Your Choices and Spending Habits (and What to Do About It). Retrieved March 20, 2017, from http://lifehacker.com/5824328/how-advertising-manipulates-your-choices-and-spending-habits-and-what-to-do-about-it

Five Key Questions Form Foundation for Media Inquiry. (n.d.). Retrieved March 22, 2017, from http://www.medialit.org/reading-room/five-key-questions-form-foundation-media-inquiry

How can I use TV and movies to teach my kids media literacy? (n.d.). Retrieved March 15, 2017, from https://www.commonsensemedia.org/news-and-media-literacy/how-can-i-use-tv-and-movies-to-teach-my-kids-media-literacy#

Kiely, E., & Robertson, L. (2016, December 19). How to Spot Fake News. Retrieved March 30, 2017, from http://www.factcheck.org/2016/11/how-to-spot-fake-news/

Knorr, C. (2012, February 21). Tips for Battling Stereotypes. Retrieved March 31, 2017, from https://www.commonsensemedia.org/blog/tips-for-battling-stereotypes

Mcclain, J. (n.d.). A Framework for Using Popular Music Videos to Teach Media Literacy. Retrieved March 28, 2017, from http://journaldialogue.org/issues/a-framework-for-using-popular-music-videos-to-teach-media-literacy/

Media. (n.d.). Retrieved March 23, 2017, from http://www.dictionary.com/browse/media

https://medialiteracynow.org/wp-content/uploads/2014/01/Drawing-the-connection-between-Media-Literacy-and-Public-Health.pdf

https://medialiteracynow.org/wp-content/uploads/2014/02/workshop-infographic-booklet-1.pdf

http://mediasmarts.ca/sites/mediasmarts/files/pdfs/tipsheet/TipSheet_Talking_Kids_Gender_Stereotypes.pdf

NAMLE Core Principles of MLE.pdf. (n.d.). Retrieved March 25, 2017, from https://drive.google.com/file/d/0B8j2T8jHrlgCYXVHSVJidWtmbmc/view

Satire. (n.d.). Retrieved June 20, 2017, from http://www.dictionary.com/browse/satire

Shapiro, J. (2015, August 31). 3 Questions Parents Of Gamer-Kids Should Ask Themselves Immediately. Retrieved March 27, 2017, from https://www.forbes.com/sites/jordanshapiro/2015/08/31/3-questions-parents-of-gamer-kids-should-ask-themselves-immediately/#3ee27bce1566

Study finds Disney Princess culture magnifies stereotypes in young girls. (n.d.). Retrieved March 25, 2017, from https://phys.org/news/2016-06-disney-princess-culture-magnifies-stereotypes.html

Talking to Kids about the News - Tip Sheet. (n.d.). Retrieved March 29, 2017, from http://mediasmarts.ca/tipsheet/talking-kids-about-news-tip-sheet

TEN QUESTIONS FOR FAKE NEWS DETECTION. (n.d.). Retrieved November 14, 2017, from http://www.thenewsliteracyproject.org/sites/default/files/GO-TenQuestionsForFakeNewsFINAL.pdf

Thoman, E., & Jolls, T. (2008). Literacy for the 21st century: an overview and orientation guide to media literacy education. Santa Monica, CA: Center for Media Literacy.

Violence. (n.d.). Retrieved June 10, 2017, from https://www.merriam-webster.com/dictionary/violence